GODLEY MIDDLE SCHOOL LIBRARY

DATE DUE			

THE INCREDIBLE WORLD OF PLANTS

THE MYSTERIOUS JUNGLES

CHELSEA HOUSE PUBLISHERS
New York • Philadelphia

Text: Andreu Llamas
Illustrations: Luis Rizo

Las Selvas Misteriosas © Copyright EDICIONES ESTE, S. A., 1995
Barcelona, Spain.

The Mysterious Jungles copyright © 1996 by Chelsea House Publishers, a division of Main Line Book Co. All rights reserved.

1 3 5 7 9 8 6 4 2

Library of Congress Cataloging-in-Publication Data

Llamas, Andreu.
  [Selvas misteriosas. English]
  The mysterious jungles / [text, Andreu Llamas : illustrations, Luis Rizo].
    p. cm. — (The Incredible world of plants)
  Includes index.
  Summary: Describes the ecology of the world's jungles, concentrating on the life forms that are found there.
  ISBN 0-7910-3465-8. — ISBN 0-7910-3471-2 (pbk.)
  1. Jungle ecology—Juvenile literature. 2. Jungle plants—Juvenile literature. [1. Jungles. 2. Jungle ecology. 3. Jungle plants. 4. Ecology.] I. Rizo, Luis, ill. II. Title. III. Series: Llamas, Andreu. Incredible world of plants.
  QH541.5.J8L5813   1996   95-15155
  574.5'2642—dc20   CIP
                                                       AC

Contents

The Fight for Light	4
Jungle Lodging	6
Jungle Clearings	8
Pollination	10
The Jungle Floor	12
Carnivorous Plants	14
Night in the Jungle	16
Epiphytes	18
Spreading Seeds	20
The Plague	22
The Strangler	24
Parasite Plants	26
Underwater Forests	28
Chemical Warfare	30
Glossary	32
Index	32

THE FIGHT FOR LIGHT

In the jungles, the fight for light is incredible. Thousands of plants and trees struggle to be taller than the rest so they can reach the sun, and some attain heights of over 200 feet (60 meters)!

In this strange competition to rise above the others, the higher one ventures, the more noticeable are the mysterious climbing plants and lianas. There are many climbing plants around the world, but most grow in the tropical rain forests. In their fight for light, climbing plants have developed different systems to achieve height. The simplest system is to hold on to a spiked stalk.

Other climbing plants produce two different kinds of roots: short and long. The short roots come out of the stalk and have hairs to help them stick to the surface that they are climbing. The long roots hang from branches and are used for getting food. Another method that plants use to climb is to launch tendrils from the leaves and stalks. These tendrils swing until they find somewhere on to which they can hold.

(1) A vegetable labyrinth
There are so many trees in the jungle that only 2 percent of the sunlight that illuminates the tops of the trees reaches the ground. This is why the jungle is dark, hot, and damp.

(2) In search of light
Climbing plants have only one goal—to grow up toward the sunlight.

(3) The ideal hideout
The mamba is one of the most dangerous snakes in the world because its poison is extremely powerful. In the jungle, the number of climbing plants and lianas provides the snakes with ideal paths for moving without being seen.

(4) Good material
Rattans are strange climbing palms which have many different uses: in wicker furniture, fish traps, hammocks, carpets, ropes, and sticks.

JUNGLE LODGING

The jungle is like an apartment building. Different kinds of animals and plants live as neighbors on each floor.

If we looked closely, we would see that plants are situated on different levels. On each level there are different conditions of light and humidity so the animals that live on each tier are also different.

The upper level is the treetops, which may be 200 feet (60 meters) high. This is the jungle roof; it prevents sunlight from reaching the ground and retains the humidity. Most animals live in the upper treetops where there are more fruits, flowers, seeds, and leaves, to feed on.

The second floor is made up of shorter trees, such as palms, which grow to 65 or 100 feet (20 or 30 meters). These trees have large, long leaves to absorb all the light they can. Many animals prefer to live on this level because the leaves are bigger and less poisonous.

The jungle floor is warm, humid, and surrounded by shade, so very few plants can grow there. Moreover, dead plants and animal carcasses rot very quickly as conditions of heat and humidity facilitate the growth of *fungi* and *bacteria*.

(1) The green house
The jungle roof is made up of extremely tall trees, and is the home of many birds (1a). On the second floor there are trees with large leaves and many other animals (1b). On the ground there are different types of lichen and flowers but not the great variety that is found on upper levels (1c).

(2) A good viewpoint
Some large birds of prey, such as the harpy eagle in South America, build their nests in the highest of the jungle tops from where they can watch their prey.

(3) Do not move
The sloth is one of the strangest inhabitants of the South American jungles. To survive it tries to pass unnoticed and so it has become the slowest of all vertebrates!

(4) Safe on the floor
Large monkeys, such as orangutans and gorillas, live on the lowest jungle level because the branches on the treetops cannot hold their weight.

JUNGLE CLEARINGS

Did you know that old trees occasionally fall down? When this happens, a jungle clearing is formed.

When one tree falls, it knocks down more trees and plants with it due to the number of lianas connecting them. This is why clearings are usually large.

Why do trees fall down? Sometimes they fall because of lightning or strong winds, but most fall when they are too old.

Temperatures are higher and humidity is lower in the clearings than in the surrounding jungle, and the death of fallen trees gives *nutrients* to the soil.

When a catastrophe, such as a hurricane or an earthquake, happens much larger clearings appear that are *colonized* by tree species, such as the cecropia, which are less than 65 feet (20 meters) tall and only live between 30 and 80 years. These trees are very important for the regeneration of the forest, because they protect the soil.

(1) A clearing in the thick
Conditions in clearings are very different from the rest of the jungle because sunlight reaches the floor and the air is much cooler.

(2) A short life
Every cecropia produces some 900,000 seeds, which scatter all over the floor. When a clearing is formed these trees grow very fast without worrying about their defense (they have no hard wood or toxic leaves).

(3) Gardener ants
The cecropia and the aztec ants have become partners. The tree gives the ants lodging and food, and in exchange, the ants protect the tree by killing the stems of the climbing plants.

(4) Retreating into the jungle
The tapir approaches the clearing from time to time; the special shape of its body allows it to break the impenetrable wall of vegetation.

(5) Eating the fallen trunks
Ant-eating bears, such as this tamandua, explore rotten trunks where ant nests abound.

POLLINATION

In the jungle the animals do the job of carrying the pollen so plants may reproduce.

There is very little wind inside the jungle, and trees of the same species are very far apart; this is why animals are so important in plant reproduction. In other *ecosystems* it is the wind that carries the pollen of most plants. To attract the pollinators, plants have flowers that produce sweet nectar, a tasty substance full of energy that animals usually get by going into the plant's *corolla*. Plants then take the opportunity to cover the animals with pollen, their sexual dust. *Pollination* occurs when the animal takes the pollen from one plant to another.

The animals that feed on nectar are a very specialized group of small animals such as insects, birds, bats, and even small rodents. To attract their pollinators, plants advertise their nectar with bright colors: yellow and white for animals with short tongues and red, blue, and violet for those with longer tongues.

The flower only gives access to animals which are good, active pollinators.

Plants that prefer insects as pollinators use flowers with *ultraviolet* coloring, which is only visible to insects.

Living helicopter
Each day the hummingbird eats half of its weight in sugar. To remain in the air it can flap its wings 80 times per second.

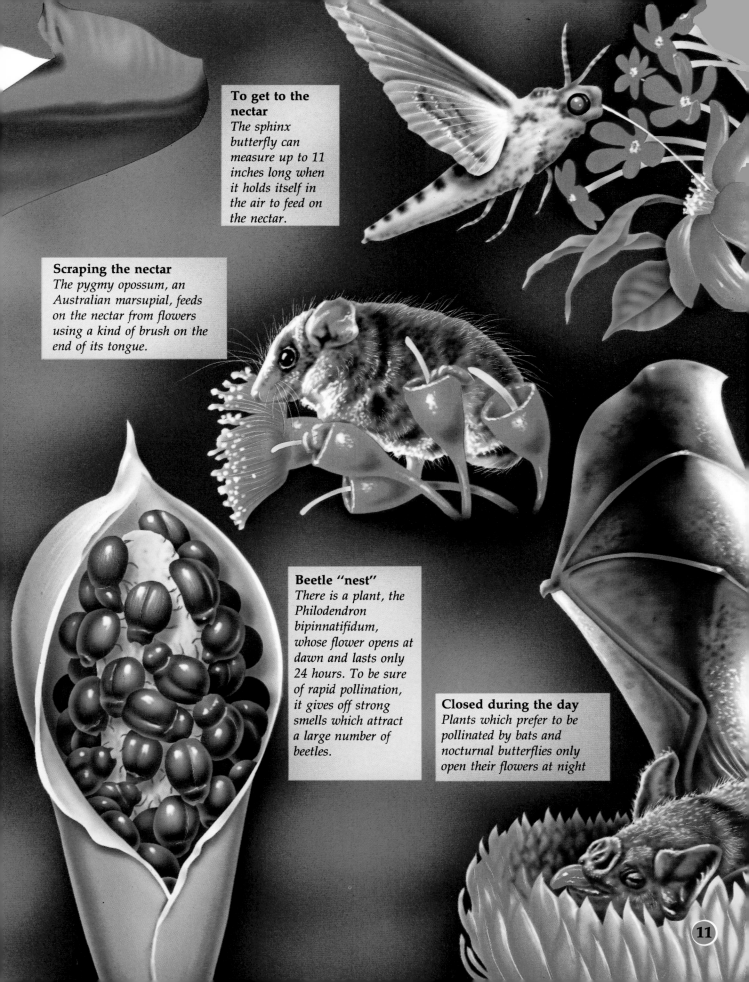

To get to the nectar
The sphinx butterfly can measure up to 11 inches long when it holds itself in the air to feed on the nectar.

Scraping the nectar
The pygmy opossum, an Australian marsupial, feeds on the nectar from flowers using a kind of brush on the end of its tongue.

Beetle "nest"
There is a plant, the Philodendron bipinnatifidum, whose flower opens at dawn and lasts only 24 hours. To be sure of rapid pollination, it gives off strong smells which attract a large number of beetles.

Closed during the day
Plants which prefer to be pollinated by bats and nocturnal butterflies only open their flowers at night

THE JUNGLE FLOOR

Although it does not seem possible, the jungle floor is poor in minerals and nutrients that are necessary for plants to grow.

In the jungle the trees themselves store most of the nutrients in the ecosystem and also extract the nutrients from fallen leaves and other forest waste. The result is that most nutrients are quickly assimilated and few are left to be swept away by the rains.

Jungle trees have adapted to living this way, and they have a spongy cloak of intertwined roots above the ground that intercept the nutrients before they escape.

There are even some roots that grow upward and enter a layer of dry leaves in order to collect leaves and fruit that have recently fallen. Also, the tree roots are covered with fungi that stick to the rotting leaves and help to direct the nutrients directly to the roots.

On the floor of the jungle clearings, where grass and bushes grow, large mammals such as tapirs, deer, and antelopes feed. These animals are the main menu of the large hunters in the jungle, such as the Asiatic tiger and the American jaguar.

The atmospheric conditions on the jungle floor are very special: there is no breeze, humidity is 90 percent, and the temperature is 83 degrees Fahrenheit. That is why it is so tiring to walk in the jungle!

(1) Deep roots
The jungle floor is covered by a large number of roots, but because the soil layer is soft, some trees strengthen themselves with roots that descend 16 feet (5 meters) underground.

(2) Fierce guards
There is a kind of bumblebee that builds its hives under the ground, covering it with a great dome of leaves, twigs, and plant stalks. There may be up to five bees guarding the hive, and they will fiercely attack any animal that approaches.

(3) Jungle hunters
The great hunters of the jungle are the Asiatic tiger, like the one shown here, and the American jaguar. The stripes and spots on their skin help them to hide.

(4) From the largest to the smallest
Some of the smallest mammals in existence live in the jungle, such as the royal antelope, which is 1 foot (30 centimeters) in height. On the other hand, some amphibians in the jungle are gigantic, like the goliath frog which can reach a length of 31 inches (80 centimeters) and weigh 33 pounds (15 kilos).

(5) Floor plants
Thanks to the damp environment, moss, lichen, fungi, and toadstools abound.

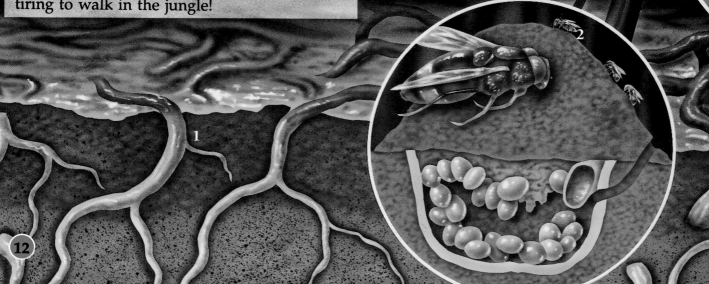

CARNIVOROUS PLANTS

Have you ever heard of the terrible carnivorous plants? Actually, they are not giant man-eaters but are mortal traps for confident insects that approach them.

Carnivorous plants live in places with nutrient-starved soil. This is why they must complete their diet to survive in a similar way to taking vitamin tablets. In any case, you must not forget that carnivorous plants also produce food through their leaves by means of *photosynthesis*.

Plants obviously cannot run after their prey, so they have learned to hunt with traps that attract small insects, mynas, and crustacea. Sometimes, small vertebrates fall into the traps (frogs, for example), but they usually get out again thanks to their strong muscles.

To attract their victims, the plants have very attractive shapes and colors and also secrete sweet, sticky, scented juices that are an irresistible temptation.

One of the most terrible is the nepenthe, which has a basin over 1 foot (30 centimeters) high, usually full of the insects it digests. Like other climbing plants, the nepenthe climbs upward to look for light, clinging to surrounding vegetation with whips that grow on the end of new leaves.

As they develop, the leaves swell and create the mortal insect traps.

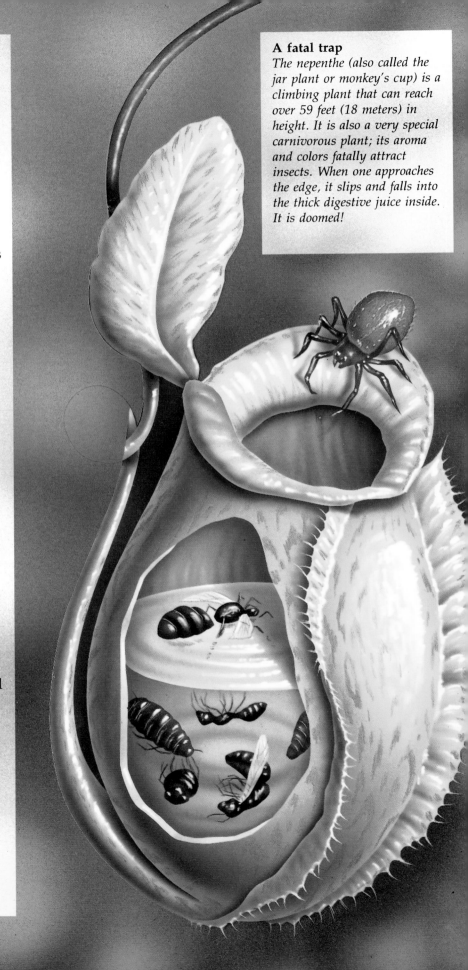

A fatal trap
The nepenthe (also called the jar plant or monkey's cup) is a climbing plant that can reach over 59 feet (18 meters) in height. It is also a very special carnivorous plant; its aroma and colors fatally attract insects. When one approaches the edge, it slips and falls into the thick digestive juice inside. It is doomed!

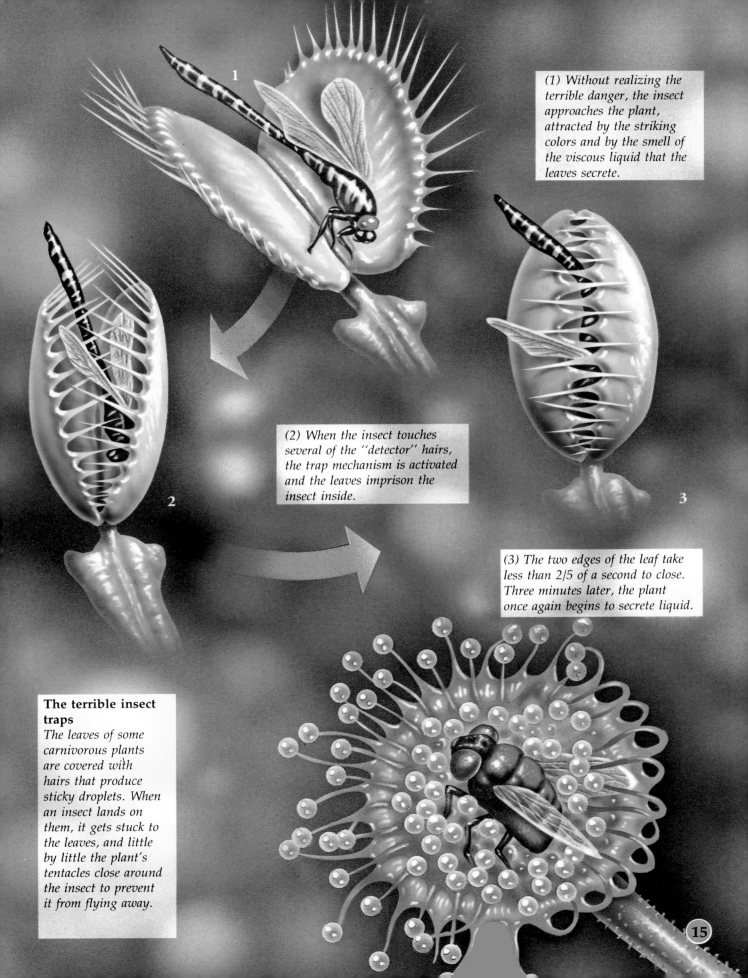

(1) Without realizing the terrible danger, the insect approaches the plant, attracted by the striking colors and by the smell of the viscous liquid that the leaves secrete.

(2) When the insect touches several of the "detector" hairs, the trap mechanism is activated and the leaves imprison the insect inside.

(3) The two edges of the leaf take less than 2/5 of a second to close. Three minutes later, the plant once again begins to secrete liquid.

The terrible insect traps
The leaves of some carnivorous plants are covered with hairs that produce sticky droplets. When an insect lands on them, it gets stuck to the leaves, and little by little the plant's tentacles close around the insect to prevent it from flying away.

NIGHT IN THE JUNGLE

Would you like to spend a night in the jungle? At night, the jungle is full of mysteries with thousands of indescribable sounds. For many of the animals and plants it is time to rest, but other species begin their activity at this time.

Many jungle plants prefer to be pollinated by bats and nocturnal butterflies, so they have gotten used to opening their flowers at night. These flowers are usually pale because during the night flowers cannot attract their pollinators with colors since there is no light. This is why they prefer to pollinate with combinations of irresistible aromas that can be smelled from a great distance.

Night in the jungle brings unexpected surprises. If one night you are a traveler lost in the jungle, you would be able to check your compass or see your watch thanks to the flat green or blue light that some fungi give off. Scientists have discovered that the luminescence of these fungi increases with oxygen and that it is related to the plant's breathing processes. These processes are particularly important at night, but why do these plants give off light?

There are also some species of insects that are capable of producing light. In the main picture you can see the luminous flight of thousands of lamp coleopterons among the trees. These insects use their luminous sparkle to mark their territory.

Night flowers
The flowers of crescentia cujete only open at night and then they give off an intense aroma similar to that of their pollinators, bats.

Luminous plants
There are fungi that shine at night with bluish green or yellow colors. Scientists believe that they use this light to attract insects.

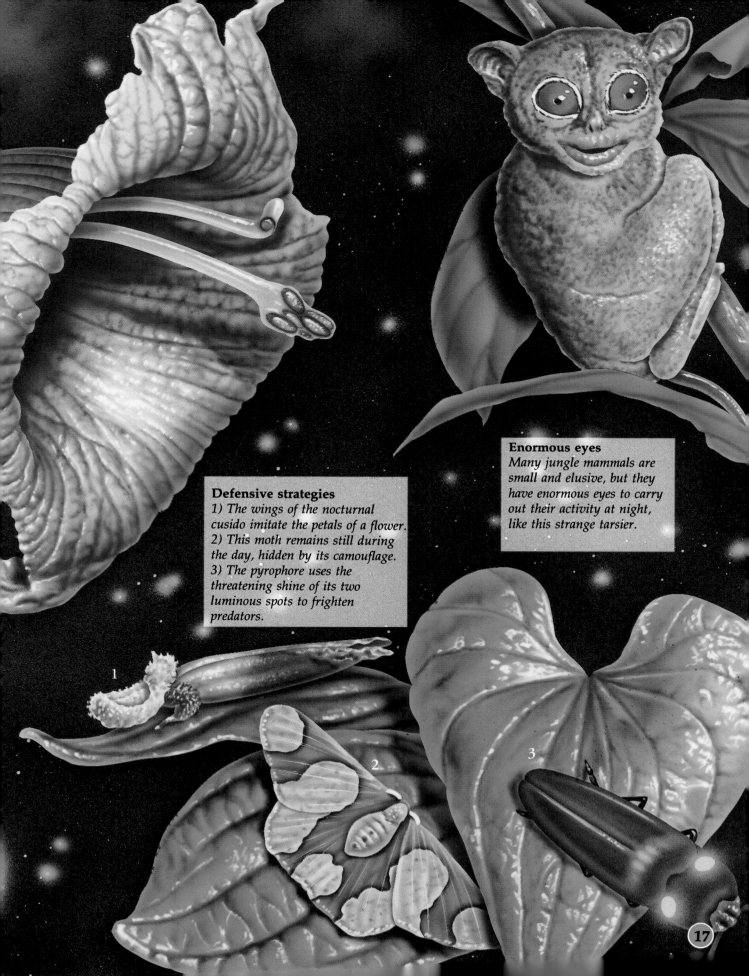

Defensive strategies
1) The wings of the nocturnal cusido imitate the petals of a flower.
2) This moth remains still during the day, hidden by its camouflage.
3) The pyrophore uses the threatening shine of its two luminous spots to frighten predators.

Enormous eyes
Many jungle mammals are small and elusive, but they have enormous eyes to carry out their activity at night, like this strange tarsier.

EPIPHYTES

Did you know that there may be up to 47 different species of orchid on one tree?

Plants, such as some orchids, which live on other plants, are known as *epiphytes*.

Epiphytes are not parasites because their roots do not penetrate the plants that support them. Actually, epiphytes grow on other bigger plants (trees, for example) without causing them any harm.

If you strolled through the jungle, you would discover that more than a quarter of all existing plants are epiphytes.

To get the water they need to live, epiphytes can absorb it directly from the air as other plants do.

However, they have their own system to supply themselves with water—they have special folds in the shape of funnels to catch rainwater.

They obtain mineral substances from the remains of other plants and from the rainwater.

(1) Paths among the treetops
With the continuous passing of animals, such as monkeys, paths are formed between the treetops.

(2) Strange partnerships
This plant gives ants lodging and nectar while the ants help to spread the seeds of the plant and protect it from invasion from other epiphytes and animals. These ants are very aggressive toward anything that approaches their partner.

(3) Frogs' swimming baths
Some epiphytic plants can hold over 5 quarts of water; certain tadpoles spend their whole infancy in these small swimming pools.

(4) Nectar thieves
Hummingbirds may peck at the base of the flower of some epiphytes to get the nectar.

(5) Rubbish disposal epiphytes
The leaves of many epiphytes have a very special shape, like a funnel. In this way they can collect the rainwater, dew, and remains of other plants.

SPREADING SEEDS

In the jungle there is hardly any wind, so plants have had to invent other ways to spread their seeds and germinate.

First, plants produce an enormous number of seeds and then produce appetizing fruit to attract the animals that will spread the seeds. These seeds are in the fruit so that when the animal eats it also eats the seeds, although it does not digest them.

Later, when the animal passes the seeds in excrement a long way from where it ate the fruit, some of these seeds will germinate and produce a new plant.

The best animals for spreading the seeds are those which travel great distances across the jungle, as the seeds can travel much further from the plant that produced them. The best seed-spreaders in the jungle are monkeys and large birds such as toucans.

To cope with the problem of hungry seed thieves, some plants cover their seeds with extremely hard shells; however, there are some animals like peccaries and agoutis that have developed ways to break the hardest shells.

The strangest fact is that some of the trees that have developed these armored shells depend on their *predators* for survival, since they are the only ones capable of opening the shell and letting the seeds out. The agouti, for instance, does not eat all the seeds because it prefers to bury and store some. Later, it does not dig up all the seeds it has buried, so some survive and germinate.

(1) No wind
The vegetation in the jungle is so dense that the air does not move. This is why plants have to turn to animals for pollination.

(2) Vegetarian bat
This is a fruit bat that feeds by absorbing the juice of jungle fruit. To eat, it puts its lips around the fruit and sinks its teeth into the flesh.

(3) Fruit near the ground
Some fruits, such as these prickly grapes, hang near the base of the trunk, very close to the ground. This is because their flowers attack insects scurrying on the ground as well as low-flying animals.

(4) Explosion of color
The wild fruit of the ginger plant have spectacular colors which irresistibly attract animals that will spread their seeds.

(5) Excessive defense
Some trees protect their fruit with such hard shells that they need help to open them. The agouti is essential for the trees' reproduction, as it opens the shell to eat the seeds. It does not eat them all but buries some, which helps with the reproduction of the trees.

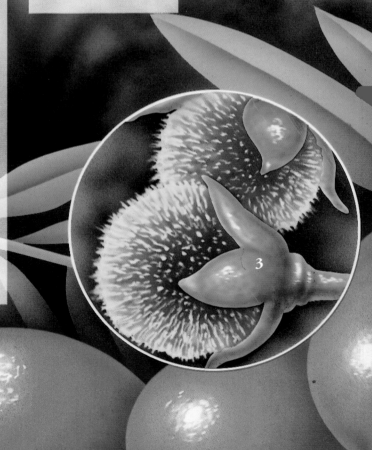

THE PLAGUE

Of all jungle animals, the most terribly destructive are the carnivorous soldier ants, also known as legionary ants.

They are feared everywhere because they advance unceasingly in columns of hundreds of thousands, attacking and devouring everything they find. This is the marabunta, the plague of ants. The appearance of a column of ants sows panic among all jungle animals. Insects with wings quickly take flight and those that jump try to leap away as fast as they can.

However, animals that are not fast enough, such as tarantulas or scorpions, may be caught and torn apart. The legionary ants are so fierce that they even enter other ants' nests, where they fight and devour them pitilessly.

(1) A powerful army
The army's march is unceasing, with its giant soldiers in front and to the sides, ready to use their jaws at any time.

(2) Taking the opportunity
The ant-eating bird carefully watches the army's march to devour some of the insects that try to escape from the ants. Then it also feeds on the remains of animals killed in the battle.

(3) The hunted hunter
The osprey is a hunting bird that glides over areas where there are legionary ants, hoping to catch an ant-eating bird.

(4) Inside the ants' nest
Every night the ants join their bodies with one another, intertwining their legs to form an extraordinary ants' nest. It is a very complicated nest with many chambers for the queen and larvae.

(5) Leaf cutters
Leaf-cutting ants form groups of more than a half million that can devastate crops. In one night they can rid a tree of all its leaves.

THE STRANGLER

The fight for life in the jungle is so intense that there is a kind of fig tree capable of "strangling" trees to take their place.

In the jungle there are many different species of strangling fig trees. Their incredible way of life is an example of the ability of plants to adapt and struggle.

The flowers of the fig tree are inside the fig itself, and they are pollinated by such specialized wasps that each species of fig tree has its own specific pollinating wasp.

Fig trees are also very fertile, and as many as 100,000 fruits have been found on one tree. The small seeds of the fig tree are spread by an innumerable amount of birds and insects that feed on the fruit. Sometimes, one of these seeds falls on a tree branch and germinates. First a long root appears in the air, and when it reaches the ground the young fig tree begins to grow fast, launching new roots to the ground and developing stalks and leaves. With time, the host plant is completely covered by the foliage of the fig tree and its trunk is caught by roots that strangle it little by little until it dies.

This is how the fig tree avoids the competition that develops on the ground by usurping the place of a tree that was already established.

1. A bird drops a fig seed on a tree branch.
2. The seed develops an air root that searches for the ground.
3. When the root reaches the ground it starts to drink water, feed, and grow faster.
4. The fig tree gradually takes the tree's light, water, and food, leaving it to die. Only the fig tree remains.

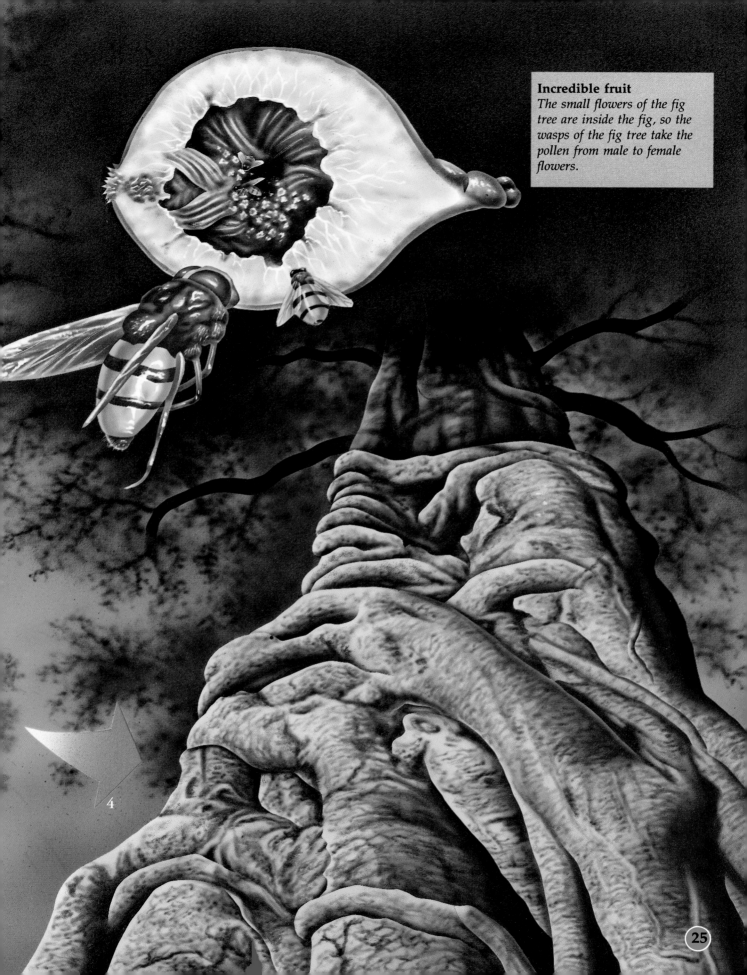

Incredible fruit
The small flowers of the fig tree are inside the fig, so the wasps of the fig tree take the pollen from male to female flowers.

PARASITE PLANTS

There are many jungle plants that prefer to steal their food instead of making it; these are the *parasite* plants.

These plants do not use the energy of the sun to produce food but have rather developed systems to steal food made by other plants, which are given the name of *guests*.

Parasite plants do not need light and many live in dark, hidden places. They do not have chlorophyll so they cannot carry out photosynthesis and therefore have no need of their leaves, which are usually very small.

To get food, a parasite plant sticks to the roots or stalk of the guest plant by means of suckers, called haustoria. These haustoria force their way into the channels along which the guest plant's food passes and they absorb the minerals and sugars.

There are some plants, such as mistletoe, that are only partially parasitic, as they also have green leaves and make part of the food they need themselves.

(1) Thousands of parasites
The fight for survival in the jungle has caused the appearance of thousands of species of parasite plants.

(2) Stealing food
Here you can see how the haustoria of the parasite plant enters the stem of the guest to steal food.

(3) Look for support
The stems of the cascade roll around the guest. Young cuscutas have roots that help them to stay on the ground. When they have grown, the roots dry and the plant hangs entirely from the guest.

(4) The biggest flower in the world
The rafflesia is a huge parasite plant that lives in the jungles on the roots of climbing plants. Its flower is the heaviest in the world at 15 pounds (7 kilos) and it has a diameter of 3 feet (1 meter)! This giant flower smells rotten to attract flies.

(5) Multicolor plants
Parasite plants have no chlorophyll so they cannot carry out photosynthesis. This is why they are not green but instead have surprising colors of red, yellow, orange, and blue.

(6) Deceiving birds
There are more than 1,000 different species of parasitic mistletoe in the tropical rain forests. The one in the picture has red flowers to attract the birds that the plant needs for pollination.

UNDERWATER FORESTS

Can you imagine fish swimming among the treetops? Well, this incredible scene is possible in the tropical rain forest.

One of the strangest scenes in the regions of the Amazon River is the enormous extension of flooded boggy forests which may be submerged below up to 39 feet (12 meters) of water for eleven months each year.

Many animals have adapted well to this situation, and there is an enormous number of fish that feed on the seeds and fruit that fall from submerged trees, giving rise to the amazing scene of fish swimming among the tops of the trees. Most plants in the flooded forest flower during the short dry season, when the water is its lowest.

Moreover, small plants may be submerged almost all year long, so they have very little time out of water.

It is precisely in this short time that plants flower, give fruit, and get as much sunlight as possible before once again being covered with water.

You must not forget these trees do not die even when they are covered with water. Life continues under the surface and some young trees may spend the first 20 years of their lives without seeing the sun.

(1) An underwater forest
In some areas of the South American jungle it rains so much that the Amazon River overflows for more than half the year. The area of forest that is underwater is larger than New York State!

(2) Fish that eat seeds
This tambaqui can smell when the rubber trees are going to produce their seeds. It then swims to the surface and eats the seeds that fall. These fish can carry the seeds great distances before expelling them.

(3) Vegetarian piranha
There are many species of piranha. This one is a vegetarian that eats the seeds and fruit that fall in the water. Other piranhas are hunters that attack their prey in groups of hundreds and can even devour a whole animal!

(4) Opportunist hunter
Some animals take advantage of the floods in the jungle. This is why, when the flooded areas emerge, jaguars come down from the branches where they have climbed and quietly hunt alligators and the fish that have been stranded on land. They are certain to get food.

CHEMICAL WARFARE

One of the most effective ways that plants have to protect themselves is producing poison.

You have seen that leaves and seeds are rich sources of food for many animals. If plants did not have systems to protect their leaves and seeds, none would survive, and with time the plant would become extinct.

Many plants, for instance, coat their seeds with poison, and the animals that eat them have to neutralize the poison before they digest them. Most jungle plants also have poison in their leaves.

Plants of different species produce different poisons, so if an animal develops defenses against the poison of one kind of plant, it will not be able to defend itself against others.

To be able to feed on plants, animals have developed various strategies. The large animals, such as monkeys, usually take small bites from different kinds of plants so they only take small quantities of various poisons. By contrast, smaller animals such as beetles are adapted to eat only one specific type of plant. The most surprising fact is that some insects are not only resistant to poison but also store it in their bodies as a system of defense against their own enemies.

Some plants have a substance in their nectar which is poisonous for all species except for their favorite pollinators.

Storing poison
These blue caterpillars are highly toxic, although the poison is not produced in their bodies. Actually, some caterpillars feed on a poisonous plant and store the poison in their bodies to later defend themselves against predators such as birds.

Poison in the skin
Some small frogs in the South American jungles have powerful poison in their skin. Indigenous people use it to poison the tips of their arrows.

Armored stomachs
The monkeys best adapted to eating leaves are the colobus monkeys of the African tropical rain forests. Their stomachs produce juices that counter the effect of the poison in the leaves when they eat them.

The good and the bad
There is a very special relationship between butterflies and plants. Plants need butterflies for pollination but the young caterpillars are usually great leaf-eaters.

GLOSSARY

bacteria single-cell microorganisms, of which there are many different species; some cause illnesses and others participate in fermentation

colony animals, plants, or places occupied by certain species

corolla the second inner circle of a flower, made up of transformed, generally colored, leaves called petals

ecosystem unit of the environment where a community of organisms live

epiphytes plants that derive moisture and nutrients from the air and rain and usually grow on other plants

fungi plants that have no chlorophyll and therefore do not feed off their own products; they usually live off decomposing organic material

guests plants that house a parasite

nutrients substances needed by a living organism for growth

parasite an organism that feeds off another organism called a guest

photosynthesis a process in which green plants synthesize organic material through carbon dioxide, using sunlight as energy

pollination the carrying or passing of pollen from the anthers (male plant sexual organ) to the stigma (female plant sexual organ)

predator an animal which, to feed itself, must stalk and attack other animals

ultraviolet a color whose radiation has a wavelength beyond that of violet on the solar spectrum

INDEX

Agouti, 20
American jaguar, 12, 28
Ants, 22
Asiatic tiger, 12
Aztec ants, 8
Bacteria, 6
Beetle, 11, 31
Bumblebee, 12
Butterfly, 11, 16, 31
Carnivorous plants, 14, 15
Caterpiller, 30, 31
Cecropia, 8
Chlorophyll, 26
Climate, 6, 12
Climbing plants, 4, 8, 14, 26
Coleopteron, 16
Colored plants, 10, 14, 15, 16, 20,
Crescentia cujete, 16
Cuscuta, 26
Cusido, 17
Ecosystem, 10, 12
Epiphytes, 18
Fig tree, 24, 25
Fish, 28
Floor (of jungle), 6, 8, 12
Flowers, 6, 10, 11, 16, 17, 18, 24, 25, 26, 28
Frog, 12, 14, 18, 30
Fruit, 6, 12, 20, 24, 25, 28
Fungi, 6, 12, 16
Harpy eagle, 6
Haustoria, 26
Hummingbird, 10, 18
Insects, 10, 14, 15, 16, 20, 22, 24, 30
Leaves, 4, 6, 8, 12, 14, 15, 18, 22, 24, 26, 30, 31
Legionary (Soldier) ants, 22
Liana, 4, 8
Lichen, 6, 12
Luminescent plants, 16, 17
Mamba, 4
Marabunta, 22
Mistletoe, 26
Monkey, 6, 18, 30, 31
Nectar, 10, 11, 18, 30
Nepenthe, 14
Night plants, 16, 17
Nutrients, 8, 12, 14
Orchid, 18
Parasite plants, 18, 26
Peccari, 20
Philodendron bipinnatifidum, 11
Photosynthesis, 14, 26
Piranha, 28
Poison, 30, 31
Pollination, 8, 10, 11, 16, 20, 24, 26, 30, 31
Pygmy opposum, 11
Pyrophore, 17
Rafflesia, 26
Rattans, 4
Roots, 4, 12, 18, 24, 26
Royal antelope, 12
Seeds, 20, 24, 28, 30
Sloth, 6
Strangling plants, 24
Sunlight, 4, 6, 8, 14, 16, 24, 28
Tambaqui, 28
Tamunda, 8
Tapir, 8, 12
Toucan, 20
Tropical rain forests, 4, 26, 28, 31
Wasp, 24, 25